The Use of the

SLIDE RULE

The Use of the

SLIDE RULE

By ALLAN R. CULLIMORE

TABLE OF CONTENTS.

	Page
Preface	2
Accuracy and Significant Figures	3
Description of the Rule	5
Decimal Point and Reading the Scale	6
Polyphase Slide Rule	14
Principle of the Rule	6
Equivalent Ratios	7
Squares and Square Roots	9
Multiplication and Division	10
Inverted Slide	13
Forms $xy = B$, $x^2y = B$	15
Cubes	17
Cube Roots	18
Logarithms	20
Involution and Evolution (General)	20
Equations Involving Fractional Exponents	21
Trigonometric Computations	22
Variation as the Square	25
Special Forms	26
General Conditions	28
Problems	29

PREFACE TO THE FIRST EDITION.

The need for a small book of this type arose in the work of teaching the use of the slide rule to engineering and industrial students, and this Manual is a direct result of sets of notes issued to classes consisting of engineering students and men of more or less practical experience. The book is not a treatise in any sense, its aim being to develop the ideas of the operator rather than to give empirical rules. Those rules that have been given are for the purpose of training the student in the formulation of processes, and it is not intended that they shall ever be committed to memory.

The examples have been taken largely from Hydraulics and Mechanics, and while the actual field covered by specific problems is narrow, the idea has been to make them fundamental. It is hoped that these examples will serve the purpose of development better than the more specialized problems arising in different branches of engineering.

PREFACE TO THE SECOND EDITION.

The second edition is made necessary by the increasing popularity of the Polyphase type of rule, a popularity rightly based on the combined efficiency and simplicity of this type. For students in engineering and vocational schools the Polyphase slide rule is to be strongly recommended. While the general principles outlined in the first edition apply equally to the Mannheim and the Polyphase types of rule, it seems advisable to make additions to certain chapters explaining the use of the Polyphase rule when the use of this type makes for more efficient calculation.

KEUFFEL & ESSER CO. NEW YORK

THE USE OF THE SLIDE RULE.

I. ACCURACY AND SIGNIFICANT FIGURES.

It is absolutely necessary for a proper and efficient use of the slide rule that the operator should have a clear and definite idea of the conditions under which the rule may be used to advantage. Even among engineers, the idea is often expressed that the slide rule is inaccurate, because, in the hands of a reasonably expert operator, the rule will give results accurate only to 1/10 of 1%. It should be borne in mind that the rule is inaccurate in exactly the same way that a four place table of logarithms is inaccurate.

In a very large class of engineering calculations, however, results which are well within the allowable error may be computed by the slide rule. The question most often arising is whether the rule is adapted to a given problem or not. To answer this, requires a knowledge of the proper use of significant figures and a close inspection of the data of the problem, together with a knowledge of the means employed in obtaining these data.

By a significant figure we mean any figure which is significant in that it gives some real information as regards the quantity which is represented. Thus:

18700.1 has six significant figures.
13.7303 " six " "
0.0032 " two " "
13000 " two or five significant figures.

Notice that in the last case an ambiguity arises and that any one or all of the zeros may or may not be significant.

Take the problem of finding the area of a circle whose radius is 4.67 feet, measured with a steel tape. It is easily seen that the recorder of the data meant that he was unable to state the distance closer than 1/100 of a foot. In short, that he felt sure that the distance was nearer .67 of a foot, than either .66 or .68. He, therefore, recorded the result 4.67 feet, using three significant figures with an accuracy of one part in 467, or not quite 1/5 of one per cent. It should be borne in mind that the number of significant figures expresses accuracy, while the number of decimal places may or may not do so. Thus 761 millimeters is identical with .761 meters, and the decimal place shows nothing. The

engineer recording or computing data should force himself to express, by the number of significant figures in the result, the accuracy of that result. The frequent habit of carrying results to a greater number of significant figures than the data warrant comes perilously near to lying with figures; it certainly creates a wrong impression as to the accuracy of the result. Certain mathematical constants can, however, be computed to any number of significant figures; for instance π may be expressed as 3.142 or 3.141592654. On the other hand, certain physical constants are very uncertain, even in the third place. Take, for example, the weight of a cubic foot of water generally given as 62.5 lbs.; conditions of temperature and solution may easily alter the last figure of the three. With recorded data, however, the number of significant figures, as well as their character, gives very definite information. If we say that light travels 186000 miles per second, we do not mean that it covers 186000 miles to within the smallest fraction of an inch in one second of time, but that the distance covered is nearer 186000 miles than it is 185000 or 187000, and the accuracy 1/186 is expressed by three significant figures. In this particular instance, it will be noticed that the three zeros to the right of the six, may or may not be significant figures. To prevent this ambiguity it has been suggested that results like the above be written $186 \times (10)^3$ which obviates the difficulty.

If, therefore, the slide rule will consistently give results to within 1/10 of one per cent. or to one part in a thousand, we have a right to use it where three significant figures are warranted in the result. The following rule given by Holman should be rigidly observed in all cases: "If numbers are to be multiplied or divided, a given percentage error in one of them will produce the same percentage error in the result." This amounts to saying that all problems, involving data correct to three significant figures only, can be computed advantageously by means of the Slide Rule. The answer is not only near enough, but is as accurate as the data warrant. Of course, in cases when the slide rule is used as a more or less rough check on logarithmic or other calculations, these questions of accuracy do not apply. The consideration of a very simple example will serve to illustrate the rule as stated above.

Suppose we wish to compute the cubical contents of a prism of earth. Consider that the horizontal distances have been measured with a tape to the nearest 1/100 of a foot, and that the heights have been measured by a level to the nearest 1/10 of a foot, and the following dimensions recorded:

Length 101.13 ft.   Breadth 7.34 ft.   Depth 9.3 ft.

Multiplying:

$$
\begin{array}{r}
101.13 \\
7.34 \\
\hline
742.2942 \\
9.3 \\
\hline
6903.33606
\end{array}
$$

Now, if the answer be as indicated, we know the contents to within 1/100000 of a cubic foot, or better, with an accuracy of 1/7000000 of one per cent.

which is, of course, ridiculous. Suppose the depth, somewhere between 9.25 and
9.34: The correct height might be 9.25, or 9.34 but if tenths alone were expressed, it would be recorded in both cases as 9.3. We see, therefore, that the
correct result lies between the following:

$$742.2942 \qquad\qquad 742.2942$$
$$9.25 \qquad\qquad\qquad 9.34$$
$$\overline{6866.221350} \qquad\qquad \overline{6933.027828}$$

We see, then, that actually we know, only, that the result surely lies beetwen
6866.221350 and 6933.027828, but we certainly know nothing more definite
than this. If we express the answer first found as 6903.33606, we know nothing
about the last seven figures. We are sure of only the first two and the result
should have been written 6900. In the light of this we now perform the same
multiplication as follows:

$$101.13$$
$$7.34$$
$$\overline{742.0}$$
$$9.3$$
$$\overline{6900.}$$

giving the answer 6900, which is as near the true value as we can know by
the data recorded.

It will be readily seen that a knowledge of the proper number
of significant figures saves an immense amount of time in calculation. This
is true no matter what means are used in calculating, whether it be multiplication, logarithms or the slide rule. The operator should accustom himself
first to examine the data of a problem and mentally calculate the desired
accuracy of the result, as well as the approximate numerical value of that result.

II. DESCRIPTION OF THE RULE.

The usual type of engineer's slide rule (K. & E. Mannheim or Polyphase)
is of wood, faced with a white composition upon which the units are graduated
in black, and is about ten inches long. Along the center line of the rule, a slide
of wood moves easily in a longitudinal groove. Each rule is provided with an
indicator or runner of glass, marked with a hair line, which serves as a reference
line in calculating. There are four distinct scales on the face of the Mannheim
and five on the Polyphase rule, and for the sake of convenience we will *always*
call the scales A, B, C, D, beginning at the top and reading down. If the
slide be inverted (that is, turned upside down in its groove) or if it be reversed
(exposing the back of the slide), the second scale from the top will always be
referred to as B, the third as C, etc. As the two top scales are double, we will
speak of the right or left A, as the case may be. If the slide be reversed, three
scales are seen on the back of the slide; a scale of sines marked S, a scale of
tangents marked T, and a scale of equal parts in the center. Notice on the

KEUFFEL & ESSER CO. NEW YORK

back of the rule proper a piece of transparent material set in the end of the rule with a line marked upon it by which the scales can be read. In operating the rule the slide will be used in four positions: direct, inverted, reversed direct, and reversed inverted. The terms for these positions are explained above.

III. PRINCIPLE OF THE RULE.

The rule is based on the principle that the addition of the logarithms of two numbers gives the logarithm of the product of the two numbers, and that the subtraction of the logarithm of one number from the logarithm of another number gives the logarithm of the quotient obtained by dividing the second number by the first. If then we add a length on the rule to another length, and these two lengths are proportional to the logarithms of certain numbers, then the length which represents the sum of these two lengths will be proportional to the product of the two numbers. This can be shown quite simply on the rule. On the back of the slide we find in the center a scale of 500 equal spaces. Suppose we reverse and invert the slide, and bring the extreme left-hand graduations into coincidence. Set the runner to 2 on scale D, and read under the runner on the middle scale of equal spaces. We read 301. Reading on $4D$ in the same way we have 602, and on $8D$—903. The distance between 1 and 2 is therefore 301 units, and between 1 and 4 is 602 units; adding these we would have 903, which, as we have seen, corresponds to 8. Division is, of course, the reverse of this process.

In describing different settings, LA and RB will be used for the left-hand scale of A and the right-hand scale of B respectively. R alone, refers to the runner. $1LA$ would mean the left-hand index on scale A. The operation, R to $3LA$, would consist in moving the runner until the hair line on it concided with the 3 on the left-hand scale of A. $3C$ to $4LA$ would mean placing $3C$ in· such a position that $3C$ and $4LA$ are the same distance from the end of the rule; that is, both would be brought into the same straight line. This is best done by placing the runner so that the line on it is on the number on the fixed scale, and then moving the slide until the number on the slide is under the line on the runner.

IV. DECIMAL POINT AND READING THE SCALE.

Success as an operator depends upon a quick and accurate reading of the graduations, and this can only be acquired by faithful practice. It should be constantly borne in mind that there is no way to distinguish by direct reading alone the position of the decimal point. The reading on the rule would be the same for each of these numbers 1751, .1751, 17.51. The rule gives simply a succession of figures in their proper order, but without determining the decimal point. This determination always must be made independently of the actual solution of the problem in exactly the same way that the characteristics of logarithms are independently computed. Suppose the problem is to set the hair-line on the runner to 478 on Scale D (R to 478D). We see that between 4 and 5 (care should be taken not to confuse this 4 and 5 with 1.4 and 1.5 which occur further to the left), there are 20 divisions, a very long division marking 4.5, 45, or 450, whichever we please to call it. Between 450 and 500 there are 4 long marks and 5 short ones; the first long one following being 460, the next 470, the next 480, etc. Our 478 lies, then, between the second and third long mark to the right of 450; that is, between 470 and 480. Between 470 and 480 we see a smaller division marking 475, and there is no division between 475 and 480. 478 lies in this blank space and lies 3/5 or 6/10 of the distance between the marks to the right of 475. In the same way, to set on 1673 on Scale D (R to 1673D), we see 1 at the extreme left followed by a smaller 1 marking 1.1D or 1100D, then a small 2 marking 1200D, etc. 1600 is easily found, and between 1600 and 1700 are ten divisions, 1650 being marked by a long one. The next short division to the right of 1650 is 1660, the next 1670, and our number lies between 1670 and 1680; 3/10 of the distance from 1670 to 1680, to the right of 1670, lies 1673D. This procedure should be gone through over all parts of the scale until the operator is familiar with the different graduations and can set on numbers quickly and accurately. It will be seen that the value of the distance between graduations changes on different parts of the rule. Three significant figures should be read on all parts of D and C; on the left, four may be obtained. On A and B, two and sometimes three are obtainable.

In regard to the fixing of the decimal point, this is, of course, fixed for any given set of data. General rules might be given for fixing the decimal point, but it seems best in a vast majority of cases to, fix the decimal point independently of the rule. This is simple in many cases and gives the added advantage of a rough check on the work. Examples will be taken up under each class of comp.:tation. It is sufficient to consider here a very simple case: Find the square-root of 1873. This evidently has a square root lying between 10 and 100; pointing off two places to the left of the decimal point is, therefore, correct.

V. EQUIVALENT RATIOS.

One of the most important uses of the slide rule is in converting quantities from one kind of units to another. Printed on the back of the rule are a series of ratios existing between different systems of units. The time saved by using these ratios is very great, especially when a considerable number of quantities are changed.

Suppose we wish to change ounces to grams and we have 6½, 7¾, 9¼, 11¾, ounces to be expressed as grams. We see on the back of the rule that 6 ounces equals 170 grams. One ounce will equal approximately 30 grams, and this knowledge allows us to fix the decimal point easily in any case. Set the runner to 6D and put 170C to the runner; we then read 6 ounces equals 170 grams, reading ounces on D and grams immediately above on C. Notice after this setting is made that one ounce equals 28.3 grams, and that one gram equals .0353 ounces. Reading directly over 6.5, 7.75, 9.25, on D, the equivalent grams are shown on C as follows: 184, 220, 262. We cannot read

over 11.67, as no scale is there, so we shift the slide to the left, putting the right index of C where the left one stood previously and then on C, by means of the line on the runner, we read 331 grams.

Consider this example: Find the intensity of water pressure in pounds per square inch at depths of 14.33, 17.84, 19.83, 9.76 feet below the surface. On the back of the rule we find feet of water : pounds per square inch :: 60 : 26, or roughly, 2 ft. of water gives a pressure of one pound per square inch; the fixing of the decimal point is, therefore, easy. Set 26C to 60D (by means of runner). Notice that one pound per square inch pressure corresponds to head of 2.3 ft., and that 1 ft. head gives a pressure of .434 lbs. per square inch; and corresponding to 14.33 ft., 17.84 ft., 19.83 ft., and 9.76 ft., we have 6.21 lbs., 7.73 lbs., 8.60 lbs., 4.23 lbs., noticing that the slide must be shifted on the first and last reading. In this connection, it is well to notice that these settings for conversion can be made on scales A and B, and then no shifting of the slide is necessary, but the loss in accuracy in most cases prohibits the use of these scales. Instead of setting 26C to 60D we might have set 60C to 26D and read feet on C to pounds pressure per square inch on D.

Ratios in constant use in a special class of calculations are sometimes more conveniently expressed as round numbers. We see on the back of the rule the ratio of the diameter of a circle to its circumference given as 113 to 355 instead of 1 to 3.1416. The first setting is somewhat easier and in all classes of work the change in ratio is easily effected. Instead of memorizing a large number of these equivalents, it is advisable to work them out when necessary, using simply the few given on the back of the rule. Suppose we wish the equivalent of U. S. gallons of water expressed in kilograms. We get the following from the back of the rule: $\dfrac{\text{U. S. Gallons}}{\text{Pounds of water}} = \dfrac{3}{25} \cdot \dfrac{\text{Pounds}}{\text{Kilos}} = \dfrac{75.}{34}$

One gallon $= \dfrac{25}{3}$ lbs. One pound $= \dfrac{34}{75}$ Kilos. Therefore, One gallon $= \dfrac{25 \times 34}{3 \times 75}$ kilos $= \dfrac{34}{9}$ kilos, or 9 gallons $= 34$ kilos. Therefore, $\dfrac{\text{Gallons}}{\text{Kilos}} = \dfrac{9}{34}$, which is the required ratio. Consider finding the equivalent of chains in feet, by the rule, Chains: Meters :: 43 : 865; Yards : Meters :: 82 : 75; 1 Chain $= \dfrac{865}{43}$ Meters; 1 Meter $= \dfrac{82}{75}$ Yards; 1 Yard $= 3$ Feet. Therefore, 1 Chain $= \dfrac{865 \times 82 \times 3}{43 \times 75}$ Feet; Chains : Feet :: 215 : 14200, or as 1 is to 66, which is the correct answer. These examples will serve to show how any desired equivalent may be obtained. They may easily be solved by following the instructions which will be given for continued multiplication and division.

The slide rule also furnishes a ready means of converting inches into decimals of a foot, and decimals of an inch into any desired fractional part thereof; it facilitates inverse operations. To reduce the number of inches and fractions, given to the nearest ¼ in., to decimals of a foot, set 1C to 12D and over

inches on D, find feet and tenths on C. For example to find the decimal of a foot corresponding to $3\frac{1}{4}$ inches, read above $3.25D$ and get $.2708$ on C. This setting can be conveniently used when the distances are given to the nearest $\frac{1}{4}$ or perhaps $\frac{1}{8}$ in. We must, of course, remember the decimals of an inch corresponding to each eighth of an inch as follows: $.125, .250, .375, .500, .625, .750, .875$. The reduction to decimals from sixteenths is simple. Set $1C$ to $16D$ and over the number of sixteenths on D find the decimal on C. As can readily be seen, this method is applicable to any fractional part, such as a sixty-fourth.

Take the following example: Express 7 9/64 in. as a decimal of a foot. Put $1C$ to $64D$, using the left-hand index on C. Reading on C over 9 on D, we find $.141$. Set $1C$ to $12D$ and read $.595$ on C, over 7.141 on D. These examples are really special cases of a general rule. To reduce fractions to decimals, set numerator on C to denominator on D. Read on C over $1D$ to find equivalent decimal, and to reduce decimals to fractions, set decimal on C to $1D$, and corresponding numerators will be found over their denominators.

Example: Find the decimal corresponding to $\dfrac{31}{47}$. Set $31C$ to $47D$, and over $1D$ read $.66$ either $\dfrac{60}{91}$ or $\dfrac{198}{300}$. We are enabled to find reciprocals in much the same way. Example: Find the reciprocal of 3.17. Set $3.17C$ over $1D$, and read on D under $1C$. Answer is $.316$.

On the back of the rule will be noticed gauge points for different metals, with the weight of one cubic foot of the metal given. In using these ratios, it is convenient to reduce all weights to equivalent weight of iron, and then reduce by means of ratios given. Suppose we wish to make a table of weights of brass plates 12 in. on a side, varying by 32nds of an inch in thickness. Dividing $\dfrac{480}{12} = 40$, which is the weight of 1 square foot of iron, 1 in. thick; set $32C$ to $40D$, and multiply in each case by 1.09, the gauge point for brass. The best method for this setting is to set $32C$ to $40D$, R to $1.09C$. Answer $1.36D$. Then R to $2C$, $1C$ to R, R to $1.09C$. Answer $2.72\,D$, etc. $\frac{1}{32} = 1.36$; $\frac{2}{32} = 2.72$; $\frac{6}{32} = 8.16$, etc. Or we might have done as follows: $\dfrac{525}{12} = 43.7$. Set $32C$ to $43.7D$ and we have as before 1.36, 2.72, 8.16. This same idea is applicable to any section of bar, plate or structural shape. (See page 10).

VI. SQUARES AND SQUARE ROOTS.

The numbers on A are the squares of those on D. We have then a table of squares and square roots ready at hand. To find the square of any number set the runner to the number on D and read the answer on A. Example: to find the square of 4.17, set the runner to $417D$ and find under runner on A 17.4. The position of the decimal point is very evident as shown. Finding the square root is the reverse of this process. Set the runner to any number

on A and under the runner find the square root of the given number. Care must be taken to choose the proper scale on A. In finding the square root of 5, if we set the runner on $5LA$, we find, under the runner on D, 2.24, and if we use $5RA$, we find 7.07. It is evident that 2.24 is the square root of 5, and 7.07 the square root of 50. We can use this rule: If a number has an uneven number of digits to the left of the decimal point, use the left-hand scale; if an even number, the right-hand scale on A. Thus we see that under the left-hand 5 we would find the square roots of 5, 500, 50000, which would be 2.24, 22.4, 224. It is perhaps better in most cases to estimate the result mentally and set on the scale so as to approximate this result. Estimating that the square root of 5 would be 2, we can quickly see that the left-hand scale must be used to obtain an answer in the neighborhood of 2.

Suppose we wish to find the weight in lbs. per lineal foot of brass rods. We have $\dfrac{525 \times 22 \times d^2}{144 \times 7 \times 4}$ lbs. which is equal (almost) to $\dfrac{51 \times d^2}{18}$. We can solve then very easily in the following manner: Set $18B$ to $51A$, read answer on A over diameter in inches on C. In setting $18B$ to $51A$, the operation has been performed with the exception of multiplying by the variable d^2. Now d^2 on B corresponds to d on C, so multiplying by d on C is equivalent to using d^2 on B, hence the setting as given.

VII. MULTIPLICATION AND DIVISION.

Notice that if we place $1C$ to $2D$, every number on D is twice the number immediately above it on C; that is, we have formed a series of proportions, using the ratio 1 to 2. All numbers on C have this ratio to the numbers immediately below on D. Suppose that we wished to solve the proportion 3.8 : 2.18 :: x : 4.53. Set $3.8C$ to $2.18D$ (always use the runner to make this class of setting) and over $4.53D$ find $7.9C$. In making this setting, place the runner first on $2.18D$ and then slide C along until 3.8 comes under the runner; then slide the runner to $4.53D$ and read 7.9 on C, which is the answer. This gives us a method of conveniently performing the operations of multiplication and division. These operations can be performed, using the two top or the two bottom scales, as determined by the percentage accuracy required in the result. To multiply on the bottom scale, set $1C$ at a factor on D; under the other factor on C read the answer on D. Setting is made as follows: Slide C until 1 is over a factor on D, slide the runner to the other factor on C, and read the answer on D under the runner. Example: 3.26×2.83. The answer is estimated as approximately 9.0. Set $1C$ to $3.26D$ and under $2.83C$ read 9.22 on D, which is the answer. Example: 3.91×7.33. The answer is estimated to be about 28. Set $1C$ to $3.91D$, under $7.33C$ find 28.7, answer. In this second example, it will be noticed that we were forced to use the right-hand $1C$ instead of the index to the left. A very considerable amount of time will be saved if the operator decides, before trying, which index he will have to use. The following example will serve to illustrate: 13.14×16.93. The answer will lie between 100 and

300. As to using the right-hand or left-hand scale, we suggest the following rule: When the first digits of the factors are such that their product gives a number greater than ten, use the right index on C; otherwise use the left. Take as examples the following:

$$2.13 \times 3.33, \quad 2 \times 3 = 6. \quad \text{Use the left index.}$$
$$7.23 \times 4.71, \quad 7 \times 4 = 28. \quad \text{`` `` right ``}$$
$$.131 \times 4.6, \quad 1 \times 4 = 4. \quad \text{`` `` left ``}$$

If this very simple rule be kept in mind much time will be saved. It will be a little difficult in certain cases, like the following, to make a decision: 3.13×3.31. By the rule, the left-hand scale should be used, considering only the first digit of each of the factors; a closer inspection will show, however, that this is not correct, and that the right index should have been used. The operator should not try to memorize these rules, but he should appreciate their significance. The one just given, if appreciated, will save an immense amount of time. If the process of multiplication be carried out between scales A and B, and if we always use the middle index on B, the second factor will never fall outside the scale of A.

Division is the reverse of the process of multiplication. Set the divisor on C to the dividend on D, read the answer on D under 1C. (The runner should be used in setting the divisor on the dividend.) Example: Divide 313 by 8.9; the answer is estimated to lie between 30 and 40. Set 8.9C to 313D and read under 1C the answer 35.2. Notice that in the case of division only one index can appear on the scale, so that no ambiguity can possibly arise.

Attention should be called at this point to a very simple setting for finding the areas of circles. Suppose that we use scales A and B. Required the area of a circle whose diameter is 11.6 inches. We use the formula $A = \dfrac{\pi d^2}{4} = .7854 \, d^2$. (Notice that .7854 is marked on scales RA and RB by a longer line in the same way that π is marked on the same scales.) We might solve this by simple division, but the following will be found easier. R to 11.6D (see the square under the runner on A), set 1B to R, and over .7854B read the answer on A. It is seen to be 105.7. Or simpler still is the following: Reading on the back of the rule: Diameter circle : side of equal square :: 79 : 70. Set 79C to 70D, R to 11.6C, and under runner read the answer on A, 105.7. The advantage of this last method is apparent, for after the first setting of 79C to 70D the area of any circle may be found by one setting of the runner. The slide rule is primarily a time-saving device and the idea always should be to choose the setting giving the quickest solution. In this special case, the last setting will always prove the most efficient when the number of areas is greater than two or three.

In engineering work, formulæ very frequently take the form of a fraction in which the denominator and the numerator are made up of several factors.

Consider the solution of the following example: $\dfrac{142 \times 15 \times 15 \times 45000}{4 \times 16 \times 16}$; first determine the decimal point mentally. The two 16s cancel the two 15s

approximately, 4 into 142 goes 35 times, $45 \times 35 = 1000$, add three ciphers and the answer will be approximately 1600000. This process of determining the decimal point should always be followed, no matter what the problem.

It is evident that we might multiply all the factors of the numerator first and divide the result by each of the factors of the denominator in turn. Or we might alternately multiply and divide. The second method is much simpler and shorter, and an actual comparison of the two methods in this problem will show the gain in speed of the second method over the first. The first method: Set $1C$ to $1.42D$, R to $15C$, $1C$ to R, R to $15C$, $1C$ to R, R to $45C$ (note here a change of index), $4C$ to R, R to $1C$, $16C$ to R, R to $1C$, $16C$ to R, read under $1C$ the answer 1404000 on D. The second method: Set $16C$ to $142D$, R to $15C$; $16C$ to R, R to $15C$, $4C$ to R; read under the $45C$ the answer 1404000 on D. It will be seen that the work of the first method is almost double that of the second. The success of this method lies in the correct choosing of divisors. Suppose that the example had been solved as follows: $4C$ to $142D$, R to $15C$, $16C$ to R, R to $15C$, $16C$ to R, R to $1C$, $1C$ to R, read under $45C$ 1404000 on D. This process is evidently more laborious. It is not always possible to arrange divisors so as to prevent the shifting of indices. Take for example, $\dfrac{45 \times 130 \times 101}{72 \times 84}$:-

mentally, 45 into 84 twice, 2 into 130, 65 times, 72 into 6500 about 90, which is the approximate answer. Probably the best solution is $1C$ to $45D$, R to $13C$, $72C$ to R, R to $1C$, $84C$ to R, R to $1C$, $1C$ to R, read the answer on D under $101C$ as 97.7; another solution is $1C$ to $45D$, R to $13C$, $72C$ to R, R to $1C$, $1C$ to R, R to $101C$, $84C$ to R, read under $1C$ the answer 97.7 on D.

The solution of the following types is almost as simple as that of simple division: $\dfrac{a^2}{b}$, $\dfrac{a^2}{b^2}$, $\dfrac{a}{b^2}$, $\dfrac{a}{\sqrt{b}}$, $\dfrac{\sqrt{a}}{b}$, $\sqrt{\dfrac{a}{b}}$

In $\dfrac{a^2}{b}$, R to aD, bB to R.   Answer on A over $1B$.

In $\dfrac{a^2}{b^2}$, R to aD, bC to R.   Answer on A over $1B$.

In $\dfrac{a}{b}$, R to aA, bC to R.   Answer on A over $1B$.

In $\dfrac{a}{\sqrt{b}}$, R to aD, bB to R.   Answer on D under $1C$.

In $\dfrac{\sqrt{a}}{b}$, R to aA, bB to R.   Answer on D under $1C$.

In $\sqrt{\dfrac{a}{b}}$, R to aA, bB to R.   Answer on D under $1C$.

The thorough mastery of the principles involved in these settings will suggest others as simple.

VIII. THE USE OF THE RULE WITH INVERTED SLIDE.

Invert the slide so that the former scale C is adjacent to A; scale C now becomes B, and B becomes C, A and D remaining the same. If we make the indices of the slide to coincide with the indices of the stationary scale, we have, reading from scale D to scale B, a table of reciprocals. Set R to $4D$ and read scale B under R, and we see .25, the reciprocal of 4. The position of the decimal point is easily determined by inspection. Find the reciprocal of 31.5. 100/32 is about 3, so the approximate answer will be .03. Set R to $31.5D$, and reading under the runner on B, we find 318, or as we know it to be, .0318. This gives ready means for converting fractions of feet or inches into decimals, provided the numerator is unity. Set R to $64D$, reading on B under the runner .01562. Or 1/64 of an inch = .01562 in.

Multiplication and division can readily be performed on the rule with the slide inverted, the process being in each case the reverse of the process with the slide direct. Invert the slide and set R to $4D$, and $2B$ to R and under $1B$ find 8 on D. In general, then, to multiply, set the runner to one factor on D and bring the other factor on B under the runner, and look under $1B$ on D for the answer. With slide inverted, set R to $4D$, and $1B$ to R, and under $2B$ find 2 on D; to divide, therefore, set the runner to dividend on D, set $1B$ to the runner and find the answer on D under the divisor. It will be easily seen that multiplication with the slide inverted is simpler than the same process with the slide direct, and all problems involving only the product of two or more quantities should be solved with the slide inverted. In multiplying with the slide direct, we found that the question arose as to which of the indices to use. This question cannot arise with the slide inverted. The only disadvantage connected with the inverted slide is that scales B and D are separated, but this is more than offset by the absence of ambiguity in choosing indices, and by the fact that both right and left indices on the slide coincide. As an example take: $137 \times 16.3 \times 8.13$; by inspection the answer will be near 21000. Solve as follows with the slide inverted: R to $137D$, $163B$ to R, R to $1B$, $813B$ to R, and under $1B$ find the result on D as 18150.

With the slide direct we would have the following: $1C$ to $137D$, R to $163C$, $1C$ to R, note here a change of index, R to $813C$, and read result on D under R, as 18150. The advantage of the first method is obvious. Both multiplication and division with the slide inverted are necessary to the speedy solution of some of the more complicated formulæ which will be shown later, and they should, therefore, be mastered.

The inverted scale is also useful in cases of equal products of the general type $ax = bc$, where x is an unknown quantity. With the slide inverted, set $1B$ to $8D$ and, with the runner set at random, the products of the numbers under the runner on B and D will always equal 8. Consider the following example: A 16 in. pulley makes 137 R. P. M.; required the diameters of pulleys driven by the specified pulley so as to give speeds of 150-175-200-250-300 R. P. M. With inverted slide, set $16B$ to $137D$, and read the required diameters on B over the speeds on D. We see corresponding to a speed of 150 a diameter of 14.6 inches, to 175—12.5, etc. This problem depends upon the fact that the speed of a pulley (belted or meshing) in revolutions per minute multiplied by the

diameter of the pulley is a constant for a constant rim speed. We have found this constant and have divided it by the speeds as given and found the respective diameters. Notice that the indices must be interchanged between speeds of 200 and 250 R. P. M.

Consider the reduction of inches and decimals to inches and sixty-fourths with the scale inverted. Required to express 14.6 in. as inches and sixty-fourths. Set R to 6D, 64B to R, the result will be found on D under 1B, and 38 will be seen to be the nearest whole number. So the answer is 14-38/64 inches. Now set R to 5D, 64B to R, and under 1B find 32, as it should be. This setting is rather inconvenient when a great many decimals are to be changed and the following is preferable. Set R to 64D, 6B to R, under 1B find 38 as before. Set 5B to R, find 32, etc. It will be seen that we really divide the decimal by the reciprocal of 64; this second method will prove most efficient when more than two decimals are to be converted. As a further example of equal products, consider this example: Two vertical pipes are connected by a tube. The cross sectional area of pipe A is 3.42 square inches, and the cross sectional area of B is 13.61 square inches. If these pipes contain water, and the surface in A is depressed 8.19 inch, what will be the increase in height in B? We have a case of equal products as in the previous example. We might solve by considering $x = \dfrac{3.42 \times 8.19}{13.61}$ and solving with the slide direct. The answer is approximately 2 by inspection. With slide direct, set 1361C to 342D, R to 1C, 1C to R, and under 8.19C find 2.06 on D. Notice in this setting that the runner must be used to make the first and last setting and that in addition the indices must be shifted, making in all five operations. With slide inverted, set 819B to 342D, R to 1B, shift indices and under 1361B, find 2.06. If conditions of accuracy permit, we may use scales A and C when the slide is inverted; this is, of course, shorter in that the indices are not shifted. Set 819C to 342A, R to 1361C, read answer on A under R, as 2.06. We might have set R to 1361A and read answer on C, giving the same result, 2.06.

THE POLYPHASE SLIDE RULE.

Multiplication, division and continued multiplication and division are very much simplified by the use of the Polyphase rule. In general 50% saving in time can be made if the relations of C, CI and D are understood. The red scale CI obviates the necessity of inverting the slide and all problems calling for inverted slide can be readily solved with slide direct by using scale CI. The study of a typical problem in detail will serve to illustrate the method and advantage of using CI. 12.8 $\times$ 6.14 $\times$ 183.2 $\times$ 62.5. Solving : First using C and D, we have, R to 12.8D, L1C to R, R to 6.14C, R1C to R, R to 183.2C, L1C to R, R to 62.5C. Answer on D under R = 90000.

This requires seven settings. Solving : Now using C, CI, and D we have R to 12.8D, 6.14CI. to R, R to 1.83C, 62.5CI to R. Answer on D under 1C = 90000. This requires four settings. With no possible ambiguity as to right or left index. Notice R to 12.8D, 6.14CI. to R gives a multiplication answer under 1C on D, to multiply this by 1.83 we need only to put runner to 1.83, answer is then under R on D. Setting 62.5CI to R multiplied by 62.5 and the answer is on D under 1C. In general for simple multiplication it is advantageous to set two factors together and read the answer, rather than to set the index on one and the runner to the other on the slide. It is, of course, very simple to remember that the processes of multiplication and division between CI and D are just the reverse of those between C and D. In this type of rule all the advantages of both direct and inverted slide are available for any particular problem.

Run through the following: $\dfrac{326}{18.44 \times 1.68 \times 47}$ R to 32.6D, 18.44C to R, notice here why CI. can be used to advantage, and the next setting is R to 47C.1 and 16.8C to R. Answer on D, under 1C = .224. The student should not leave this subject until he sees clearly the interrelation between C D and CI.

It frequently happens that we wish to successively divide a number by various other numbers. Suppose we time 85 revolutions of a current meter in 47.3, 48.1, 46.4 seconds, respectively, and wish to reduce to revolutions per second. Scale D with CI should be used. Set 1C to 85D and read under 47.3, 48.1, 46.4 and we have 1.80, 1.77, 1.83 revolutions per second, respectively.

IX. THE FORM $XY = B$ AND $X^2Y = B$.

Let us now consider problems of the general type $ab^2 = x$; that is, to find the product of two factors, one of which is the square of a given number. Find $(3)^2 \times 9$. Set 1C to 3D, read answer on A over 9B, which is 81. It is evident that we first square 3 and multiply the answer by 9 on scales A and B. The reverse of this problem occurs in the designing of beams, the problem of design being to determine the dimensions of a section of a beam to withstand a given moment. As this problem is typical of the use of the slide rule in a wide variety of problems, it will be taken up more in detail than the problem itself might seem to warrant. Consider the design of a rectangular wooden beam to withstand a moment of 50000 in. pounds. We have from statics $M = 1/6\ fbh^2$, where $f =$ allowable stress in cross bending in pounds per square inch, b the width, and h the depth of the beam in inches. Let 1200 lbs. equal f, and we have $50000 = 200\ bh^2$ or $250 = bh^2$, and we wish to determine b and h so that their product shall be 250 or very slightly in excess of that figure. It is, of course, the exact reverse of the problem above. The problem as given is evidently indeterminate, as for every valve of b assumed we can find a corresponding value for h. There are, as a matter of fact, certain limitations. We desire a beam of minimum material, and as a commercial size must be used, we desire a beam whose theoretically desired section shall vary as little as possible from the commercial sizes. For the purposes of this problem let us say we can obtain beams 2×6, 2×8, 2×10, 4×6, 4×8, 4×10, 6×6, 6×8, 6×10.

Set R to 250A (the left-hand 250). The reason for this choice will be pointed out later. Slide left 1B to the runner. We then see that $(15.8)^2 \times 1 = 250$, or a beam 1 in. in width would require a depth of 15.8 in. Put 2B under the runner, h then would have to be greater than 11 in., but 2×10 is our

deepest commercial size. Put $4B$ under the runner and we see by the rule that $h = 7.9$, that, therefore, a 4×8 would be sufficient with a waste of .1 of an inch in height $\times$ 4 in. in width or .4 square ins. Put $6B$ to runner, $h = 6.47$, or we would require a 6 in. $\times$ 8 in., as the smallest commercial size available with a width of 6 in. We have seen that a 4×8 is sufficient; the 6×8 then is much too large. Notice that for each assumed width the rule shows the waste between the best theoretic section and the commercial section of the same width. Now as to the choice of the right or left-hand scale on A. As a matter of fact, either may be used if the proper scale on B is used. Suppose we had used LA with RB, for $b = 4$ we would have found $h = 25$. Obviously $(25)^2 \times 4$ is not equal to 250, but $(2.5)^2 \times 40 = 250$ and such a choice would theoretically fulfill the conditions of our problem. No ambiguity will arise if we roughly check our choice once for each problem.

Let us consider one more example: $1225 = bh^2$. Set R to $1225LA$, set $8LB$ to R; we find h equals approximately 40. Now $(40)^2 \times 8$ does not equal 1225. So use $8RB$ and we find $h =$ approx. 12.5 and $(12.5)^2 \times 8 =$ approx. 1225. Rules might be given for this class of calculation, but the above method is simpler and more satisfactory. Use either scale on A and determine by trial the scale to be used on B.

Consider another problem of the same type but simpler in principle. We desire to determine the number, width, and thickness of steel plates to be used in the flange of a plate girder. The plates must have a gross sectional area of 33.6 square inches. They may vary in width from 12 to 16 by half inches (that is, we may use 12, $12\frac{1}{2}$, 13, $13\frac{1}{2}$, etc.) and the plates may vary in thickness from $\frac{3}{8}$ in. to $\frac{5}{8}$ in. by 16ths of an inch (that is, we may use $\frac{3}{8}$, $\frac{7}{16}$, $\frac{1}{2}$, etc., up to $\frac{5}{8}$). Required the number of plates, their width and thickness, so that their total sectional area will be as little in excess of 33.6 in. as possible. The plates must, of course, be all the same width, but each plate may vary in thickness between the limits specified; further it is desired to have the plates of as nearly uniform thickness as possible. Set R to $33.6D$, set $12C$ to R. Read under $16C$, 44.8 on D. This means that with a width of 12 in. the plates must total $\dfrac{44.8}{16}$, but we consider only full 16ths according to the conditions of the problem, so 45/16 are required. Tabulating waste of each width from 12 to 16, sliding the runner to $12\frac{1}{2}$, 13, etc., to determine 16ths required, we have:

12	$12\frac{1}{2}$	13	$13\frac{1}{2}$	14	$14\frac{1}{2}$	15	$15\frac{1}{2}$	16
.25	0	.7	.2	.6	9	2	3	4

So $12\frac{1}{2}$ shows least waste, and putting $12\frac{1}{2}$ under runner we have 43/16 as total thickness. Now two whole numbers the product of which is nearest 43 are 7 and 6, so use five plates $\frac{7}{16} \times 12\frac{1}{2}$ and one plate $\frac{1}{2} \times 12.5$.

Checking: $\dfrac{5 \times 12.5 \times 7}{16} = 27.3$ and $\dfrac{12.5}{2} = 6.25$, and $\dfrac{27.3}{6.25}$

$$\overline{}$$
$$33.55 \ chk.$$

THE POLYPHASE SLIDE RULE.

With the Polyphase rule cases of $xy = B$ are sometimes simpler of solution using CI. in conjunction with D. Consider the problem as given above. Set 16CI to 33.6 D. Run to 12CI, read under R on D, 44.8, which is the number of 16ths required in the total thickness of plates with a wastage shown. Shift R to 12.5CI and we see 43, with no wastage. R to 13 CI under R 41.3 with .7 wastage, so 12½ width plates should be used.

X. CUBES.

Cubing a number (a) is, of course, the same as multiplying a by b^2 when (a) and (b) are equal. To cube 2.17, set R to 2.17D. Set 1B to R and over 2.17B read answer 10.2 on A. It is frequently of advantage to cube a number with the slide inverted. This setting is in reality simpler than with the slide direct and should always be used in connection with problems in which for any reason the slide is in the inverted position. To cube 2.17, set R to 2.17D, set 2.17C to R, read over 1B the answer 10.2 on A. Notice that this operation consists in squaring the number in the regular way, and then multiplying the result by the number itself; using scales A and C. It is interesting to notice that if we set R to 2.17D, 2.17B to R and read over 2.17C on A, we have the answer on A, 10.2.

The application of these last settings is clearly seen when we wish to raise a number to the 3/2 power. This can be performed very simply with the slide direct, but the solution with slide inverted is in many ways more satisfactory. Let us consider the first problem with the slide direct. To raise 3.163 to the 3/2 power. Set 1C to 3.163D and read the answer on D under 3.163B, which is 5.63. Notice if we had used the right-hand index instead of the left we would have read 1780. Suppose we perform the same operation with the slide inverted. Set R to 3.163D and 3.163C to R, read answer on B under 1A, 5.63, or on D under 1C. We will notice that two readings under (1) are possible; one reading being 5.63 and the other 1780. There will, of course, never be any doubt as to which of these two values is correct in a particular example, if the operator forms the habit of estimating the result. In this case, $(3.16)^3$ is approximately 30, and $\sqrt{30}$ lies between 5 and 6. As to the value 1780, it is, of course, $(31.63)^{\frac{1}{2}}$; the two values arising in finding the square root of 31.63 and 3.163 cubed. In making this setting, notice that under the runner on scale C is the number itself, on A is the square of 3.163, on B is $\sqrt{(3.163)^3}$. Under one of the C indices on D read $\sqrt{(3.163)^3}$ and under the other $\sqrt{3.163}$.

THE POLYPHASE SLIDE RULE.

Below D on the edge of the polyphase will be found a scale of cubes. Set R on any number on D and read the cube of that number on the edge of rule below D. It should be noticed that the accuracy obtainable by using this scale is not as great as in the method given above, but it is sufficient for many purposes, being approximately ½ of 1%.

XI. CUBE ROOTS.

In finding the cube roots of numbers, the slide may be used either direct or inverted, and in either case the process is the reverse of finding the cube with the corresponding position of the slide. We must in general recognize three cases. Take for example the problem of finding the cube roots of the following: 270, 27.0 and 2.70. These will serve to outline the process. To find $\sqrt[3]{270}$ with the scale direct: R to 270RA, slide B and C to the left, using thereby the right-hand scale on B until the same number on B appears under the runner that is seen on D under the right-hand index on C. When 6B is under R, we read 6.7 on D, and when 6.46B is under the runner, we read 6.46 on D under 1C. Therefore, 6.46 is $\sqrt[3]{270}$. To find $\sqrt[3]{27.0}$, set R to 2.70LA, and using the same method with RB, we read 3 on B under R and also 3 on D under 1C. Using RA and LB, we arrive at the same result. To find $\sqrt[3]{2.7}$, set R to 2.7LA and move slide to the right so as to use LB. 1.39 then appears as the $\sqrt[3]{2.7}$

Tabulating, the following are evident:

$\sqrt[3]{270}$ = 6.46 Using right-hand scales.

$\sqrt[3]{27.0}$ = 3.00   "   " A with left B or left A with right B.

$\sqrt[3]{2.70}$ = 1.39   "   left-hand scales.

This general rule can then be formulated: If a number is greater than unity, point off periods of three places to the left of the decimal point; and if there are no digits to the left of the last period, use the two right-hand scales; if one digit remains to the left, use the two left-hand scales; if two digits remain to the left, use the right or left with the left or right. With a number greater than unity, in most cases it is easier roughly to approximate the result and use the combinations of scales giving this result. To find the cube root of 31700 by rule, point off 31'700 and we have two digits to the left, and the rule calls for left with right or right with left. Using the former we get 31.8. The cube roots of fractions presents greater difficulty. Making a table as above we have:

$\sqrt[3]{2.7}$ = 1.39 Using the two left-hand scales.

$\sqrt[3]{27}$ = .646   "   "   " right " " "

$\sqrt[3]{.027}$ = .30   "   "   " right with the left or vice versa.

$\sqrt[3]{.0027}$ = .139   "   "   two left-hand scales.

This leads to a rule for fractions as follows: Point off the number of zeros immediately to the right of the decimal point in groups of three, counting the decimal point as one zero; if there are no zeros immediately to the right of the last group, use the two left-hand scales; if one zero remains, use the two right-hand scales; if two zeros remain, use right with left or vice versa. To find the cube root of .000000738, pointing off .00'000'0738, we have one zero remaining; so, by the rule, we use two rights and read 904, and as we have two groups, .00904 is the answer.

It should be distinctly understood that these rules are not intended to be memorized, but a study of them will simplify cube root computations. As already suggested, if the number is greater than unity, a method of approximation and trial is most satisfactory.

With fractions, if a rule be desired, the following is simple: Count off the zeros between the decimal point and the first significant figure of the number in the following way (counting the decimal point as one zero): Right, either, left, right, either, left, etc., the scales to be used being determined by the word falling on the last zero. As, for instance, in the following: .000386. Counting as indicated above, we use the two right-hand scales, and read .0728 as the cube root desired.

In finding the cube root with the slide inverted, the rules given above regarding the proper reading of the scales hold. Suppose we wish to find the cube root of 1734; evidently the result lies between ten and twenty. Set R to 1734LA and place the middle index on C to R. Find a value on C such that the number on the same point on D is identical. Reading on RC, we find immediately above 2585 on D, 2585 on C. Obviously, this cannot be the cube root sought, so reading now LC we read 1202, so 12.02 is the root sought. Again taking the same example, we might have applied the rules as given. Pointing off 1'734, one digit remains, and the two left-hand scales are called for. These are the ones which we have used and the result will be as before. Notice if the number had been 17340, we would have used, as we did in the first trial, the LA with the RC and found 25.85, which would have been correct had the number been 17340.

Take for another example the cube root of .000892; counting according to the rule given, we end on the word "right" and we find .0963 as the correct answer. In finding the cube roots with the slide inverted, the runner should be used in finding the coincidences, as it helps the eye materially and is, of course, not needed at the center index of C after the placing of that index.

We can now solve problems of the type $\sqrt[3]{(84)^2}$. With the slide direct, set R to 84D and move the slide until the same number on B is under R, that is seen on D under the left index of C; in this case, using the left scale on B, we read 413. Notice that $80^2 = 6400$ and that 41.3 cannot, therefore, be the number sought; however, if we use RB we find 192 and 19.2 is the correct answer. The question of which slide to use, direct or inverted, is largely a personal one. The operator should, however, master both methods.

THE POLYPHASE SLIDE RULE.

With the Polyphase rule cube roots may be found to an accuracy of ½ of 1% by the use of the scale on the edge of the rule. It will be noticed that the scale is divided into three parts. The left-hand scale being for numbers of one digit to the left of decimal point, the middle scale for numbers of two digits to the left and the right-hand scale for numbers of three digits to the left of the decimal point. If the number of digits be greater than three, subtract any multiple of three and use the scales as indicated above; thus, for the cube root of 1,860,000, subtract 6 from 7 and find cube root of 1.86 on scale to left, setting runner to number on the cube scale and reading the answer on D as 1.23.

It will be seen that the cube scale makes problems of the type $A^{\frac{2}{3}}$ very simple. Set R to A on the cube scale and read the answer on A under R. $10^{\frac{2}{3}} = 4.65$. Set R to 10 on cube scale and read 4.65 on A under R.

XII. LOGARITHMS.

Logarithms of numbers may be found on the slide rule in two ways. First, set $1LC$ to the number on D and obtain the mantissa of the logarithm on the center scale of equal parts to be found on the back of the slide, by reversing the rule and reading under the line etched on the xylonite on the left end of the back of the rule. To find the logarithm of 265, set $1C$ to $265D$ and reading on the center scale on the back of the rule under the line, we find 423. The desired logarithm is then 2.423. It should be remembered that the number found on the rule gives only the mantissa of the logarithm and that the characteristic must in every case be supplied in order to complete the logarithm. In the second method, we reverse and invert the slide and read the mantissa on the center scale directly over the number on D; the runner being, of course, used to do this. Thus we find the logarithm of 265 to be 2.423 as before.

In finding the numbers corresponding to logarithms, the reverse of either of these two methods is employed; the mantissa alone being set upon the scale. Example: to find the number corresponding to the logarithm 3.697: With slide direct, set 697 on center scale on back of slide to line on xylonite on back of rule, and read answer on D under $1C$, 4980; or reverse and invert the slide, put the indices in coincidence and set runner to 697 on the scale in center, and read under runner 4980, the answer, on D.

XIII. GENERAL INVOLUTION AND EVOLUTION.

Logarithms are used in slide rule work mainly to raise numbers to fractional powers. In raising numbers to integral powers, we can use either the logarithmic method or a combination of squaring and cubing with direct multiplication. Thus: $(317)^7 = [(3.17)^2]^2 (3.17)^3$ or $(3.17)^2 \times (3.17)^2 \times (3.17)^3$. This method is applicable to any power or root. Suppose we solve the problem first by the method outlined above, with this change $(3.17 \times 3.17)^2 \times (3.17)^3 = (3.17)^7$. Slide direct; $1C$ to $3.17D$, R to $3.17B$, $1B$ to R, R to $101B$. Read answer under R on A, 3210. The logarithmic method depends, of course, upon the fact that if we multiply the logarithm of the number by the exponent, the answer is the logarithm of the required number. Set $1C$ to $3.17D$. Read on center scale on back of slide .501, the characteristic being 0. Set $1C$ to $501D$, R to $7C$. Read on D under R 3.505. Set 505 on center scale on back to mark on xylonite and read under $1C$, 3210 on D, which is the answer. It may be easily seen that the first method is the more accurate, while for the seventh power, at least, the second is much the shorter. The logarithmic method is especially adapted to problems involving fractional powers or roots, except where the desired root or power is a multiple of two or three. For instance, suppose we wish to find $(717)^{\frac{3}{5}}$, or as it might be stated, the fifth root of 717 cubed. We may cube 717 and extract the fifth root. Cubing 717 gives 370000000, and the logarithm of this number is by the rule 8.568. Dividing by 5 gives 1.713 and the number corresponding to the mantissa .713 is 516, so the correct answer will be 51.6. An easier solution is this: Divide 3 by 5 giving .6; the logarithm of 717 is 2.855; $2.855 \times .6 = 1.713$ and the number

corresponding as before to the mantissa .713 is 51.6. The problem as outlined above is actually worked out as follows: Mentally $\frac{5}{3} = .6$; set $1C$ to $717D$, read under line on back of rule 855, and with characteristic 2 the logarithm is 2.855; set $1RC$ to $2.855D$ and read on D under $6C$ the number 1.713; set 713 on the center scale on the back of the slide to the line, and read on D under $1C$, 516, and as the characteristic was 1 the answer will be 51.6. Remember that the rule in every case gives only the mantissæ and that the characteristic must be supplied independently.

The problem of finding the root of numbers, the index of the root being fractional, is slightly more complicated. Solve $\sqrt[.03]{1.9}$ as an example: This is equivalent to $(1.9).^{\overline{03}}.$. The value of $1/.03$, determined by the method of finding reciprocals as previously explained, is 33.3. The logarithm of 1.9 = 0.279; $0.279 \times 33.3 = 9.3$; number corresponding to mantissa .3 is 2. So the answer will be 2 000 000 000.

As another type, take $(.00273)^{.232}$, which represents the most difficult class of problems in involution. Find logarithm of .00273 to be 7.436—10 or —3.436, remembering that the characteristic is negative and the mantissa positive. $-3.0 \times .232 = -.696$

$.436 \times .232 = \underline{.101}$

$-.595$ adding and subtracting 10, the logarithm is 9.405-10; and the number corresponding to mantissa .405 is 254. Therefore, $(.00273)^{.232}. = .254$

Let us now check this result as follows:

$(.00273)^{.232} = .254$ if the work is correct.

$(.00273)^{3}/_{13} = .254$, for we set $232C$ to $1D$ and under $3C$ find $13D$.

$(.00273)^{3}$ should equal, if the work performed is correct, $(.254)^{13}.$

$(.00273)^{3} = .0000000203.$

$(.254)^{13} = \{[(.254)^{2}]^{2}\}^{3} \times .254.$

$(.254)^{2} = .0646.$

$(.0646)^{2} = .00416.$

$(.00416)^{3} = .000000072.$

$.000000072 \times .254 = .0000000183chk.$

THE POLYPHASE SLIDE RULE

The methods of squaring and cubing having been explained, it is easy to see how to raise a number directly to the 4th, 5th and 6th power by the use of CI scale.

a^{4}, $aC1$ to aD. Answer: A over $1B$.

a^{5}, $aC1$ to aD. Answer: A over aB.

a^{6}, $aC1$ to aD. Answer, on cube scale under $1C$.

XIV. EQUATIONS INVOLVING FRACTIONAL EXPONENTS.

Equations of the type $X^{N} = b$ may be solved by the methods already given, for $X^{N} = b$ is equivalent to $X = b_{N}^{\frac{1}{N}}$. Now let us consider a few examples occurring in engineering work. The volume of the compression space in the cylinder of a gas engine is $\frac{1}{5}$ of the piston displacement. What is the efficiency of the engine? The formula to be used is $E = \left(\frac{Vd}{Vc}\right)^{k-1}$ where $Vd = \frac{1}{5} Vc$ and

$k = 1.405$. Substituting these values in the formula, $E = (.25).^{405}$. Set $1C$ to $25D$ and read under line on back of the rule on the center scale 398, so the logarithm required is -1.398, $-1 \times .405 = -.405$

$$.398 \times .405 = .1609, \quad 1RC \text{ to } .405D, \text{ read}$$

under $398C$. Next subtract

$$-.244$$

$-.244$ from 10-10, and we have, as the logarithm of the answer, 9.756-10. Set 756 on the logarithmic scale on the back of the rule and read on D under $1C$, 574; $E = 1-.574$ or about 43%

Take as another example, Hodgkinson's formula for the strength of cast-iron columns the lengths of which exceed thirty times their diameters.

The formula is $W = 99318 \dfrac{D^{3.55}}{B^{1.7}}$; where W equals the breaking load; D the diameter of the column in inches; and B its length in feet. Let $D = 3''$; and $B = 11.5'$; $W = \dfrac{3.^{3.55}}{11.5^{1.7}} \times 99318$. The example may be solved by finding the numerator and the denominator of the fractions separately and then dividing them.

Log. $3 = .477$ Log. $3^{3.55} = .477 \times 3.55 = 1.695$ Number corres. $= 49.5$
Log. $11.5 = 1.061$ Log. $11.5^{1.7} = 1.061 \times 1.7 = 1.805$ " " $= 63.8$

So $W = 99318 \times \dfrac{49.5}{63.8} = 76900$.

This method is quite simple, but the following in many cases will prove to be more satisfactory:

We may write the formula: $W = 99318 \dfrac{3^{3.55}}{(3^{3.22})^{1.7}}$. Since $3^x = 11.5$,

and $x = \dfrac{\text{Log. }11.5}{\text{Log. }3} = \dfrac{1.061}{.477} = 2.22$; then we have $\dfrac{3^{3.55} \times 99318}{3^{3.77}} = W = $

$99318 \times \dfrac{3^{3.55}}{3^{3.77}} = 99318 \times \dfrac{1}{3^{.22}}$; and since Log. $3^{.22} = .477 \times .22 = .105$ and

the number corresponding $= 1.27$, we have $W = \dfrac{99318}{1.27} = 77000$. This will cover most of the cases which arise in involution and evolution.

XV. TRIGONOMETRIC COMPUTATIONS.

Trigonometric functions may be found on the rule either with slide direct or reversed; sines and tangents being given directly by the readings on the rule. On the reverse of the slide will be found two scales, one at the top marked S giving sines, and the lower marked T giving tangents. To find the sine of an angle with the slide direct, set the angle on scale S to the line on xylonite on the back of the rule, and read the sine of the angle on B under the index on scale A. For example, to find the sine of 15°, we set 15 on scale S, to the line and under middle or right index of A we read .259; if we wish the tangent of 15°, we set 15° on scale marked T to the line on the back of the rule and read on C under index on D, .263. Or the slide may be reversed — the left-hand indices brought into coincidence, and the sines and tangents read directly, using scale of sines with A and scale of tangents with D. Reading on A over 15° on B we have .259, and reading on D under 15° on C we find .268. Notice that if the sine occurs on the scale LA, one zero stands between the decimal point and the first significant figure of the desired sine; the sine of 2° is .0349 and the sine of 20° 26' is .349. This should be kept clearly in mind. In the same way, all the values of the tangent to be read on scale D will lie between 1. and .1. The sines and tangents of small angles are practically identical, and may be found in one of two ways:- either by using the gauge point on the rule, or by an application of the proposition that the sines and tangents of very small angles are proportional to the angles themselves. Using the first method, let us find the sine of 4' and 56". Set the gauge point which is seen just before the 2° mark on the scale B to $4A$ and read on A over $1B$, 1162. Then set gauge point which is near 1° 10' to $56A$ and read on A over $1B$, 2715. Remembering that the sine of 1" is .000005 and that the sine of 1' is .0003, we will write our first reading .001162 and add to that our other reading 0.002715 and obtain .0014335, which is either the sine or tangent of 4' 56".

The second method is as follows: The sine of 4' 56" is equal to 1/10 the sine of 40' 560" = 1/10 sine 49.33'; and by the rule we find sine 49' 33" = .0143 and 1/10 of this value is .00143. Suppose we multiply 4' 56" by 20, we have sine 4' 56" = 1/20 sine 1° 38.66'. We find sine 4' 56" = 1/20 = .0286 = .00143; the latter method is preferable in almost all cases. As will be noticed, we cannot find directly the sine of angles between 90° and 360°, nor tangents of angles greater than 45°. All that is necessary is, of course, to reduce the functions as expressed to functions of an angle less than 90°. Thus the sine of 169° = sine 11° by the familiar rule of trigonometry, and the tangent of 75° = $\dfrac{1}{\tan 15°}$. In cases where we have to deal with other functions than the sine or tangent we have to transform the function by trigonometry until the required function is expressed in terms of the sine or the tangent. Cosine A = sine $(90°-A)$; that is, to find the cosine of 49° 13' we find the sine of 40° 47', which by the rule is .653. This method enables us to find all the functions very simply: $\cos A = \sin (90°-A)$, $\cot A = \dfrac{1}{\tan A}$, $\sec A = \dfrac{\tan A}{\sin A}$, $\csc A = \dfrac{1}{\sin A}$, or in finding the cosine of A we might divide the sine of A by the tangent of A. Thus to find the cos 30° 15', find .503 as the sine of 30° 15' on scale A, set R to $503D$, 30° 15' C to R and read under $1C$ .864 on D, which is the cosine required. In general, the first method is easier when the angle lies between 45° and 90°, and the second, when the angle lies between 0° and 45°. Let us consider the example of finding the cotangent of 75°; evidently cot 75° = tan 15°. Notice that with the slide reversed we can multiply directly by the sine or tangent of the angle without first determining the actual numerical value of the function; needless to say, the process of division is equally simple. Solve $\dfrac{217.3}{\tan 13° 14'}$; set 13° 14'$C$ to $217.3D$ and reading under $1C$ find on scale D 923.0 as the answer. Solve $653 \times \sin 36° 41'$; set $1B$ to $653A$, R to 36° 41'B and read answer 390 on A under R.

2.17

4.36

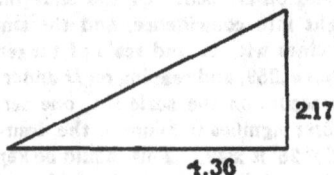

Required the solution of the right triangle as given above. With scale direct divide 2.17 by 4.36 and find .497 as the result. Let the runner rest at .497 and reverse the slide, set indices together and read on C under R, 26° 25', which is the value of the angle opposite the shorter leg; the other angle will be 63° 35'. Set R to 2.17A, 26° 25'B to R and read on A over 1B 4.86, which is the length of the hypothenuse. Check this result by squaring 2.17 = 4.71, and squaring 4.36 = 19, adding the two = 23.71, extracting the square root = 4.86 chk.

We may solve an oblique triangle using the law of sines very simply, in any case where that law applies.

Solve the following oblique triangle:

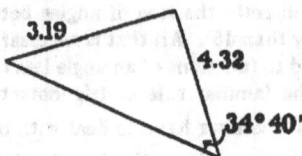

3.19

4.32

34° 40'

With the slide reversed, set 34° 40' B to 3.19A and read the answer on B under 4.32A, as 50° 35', which is the value of the angle opposite the side 4.32. The other angle will, of course, be 94° 45' and since we know sine 94° 45' = sine 85° 15', we can easily find the side opposite, by a simple proportion as above.

In general, it can be stated that any formula adapted to logarithmic computations is adapted to computations with the slide rule. If a particular problem calls for logarithmic functions, they can be very simply found by finding the logarithm of the function. Suppose log tan 38° is desired. Reverse the slide and set the runner on 38° C, then invert the slide reversed as it is, and read on the scale of logarithms under the runner .893, which is the required mantissa, the complete logarithm being 9.893-10. In finding the log sine use the same general process, but multiply the mantissa as obtained on the scale by 2, and take the fractional remainder as the required mantissa. Find the log sin 20°. With the slide reversed and the indices coinciding, set R to 20° B, invert the slide and read on the logarithmic scale under R, 767. Multiplying by 2 gives 1.534 and .534 is the mantissa required. The logarithmic functions are often useful in checking logarithmic work.

XVI. VARIATION AS THE SQUARE.

The following example is typical of a wide variety of problems to which the slide rule offers a simple solution. The example as given is very simple, but from it others may be deduced. Given the parabola as shown:

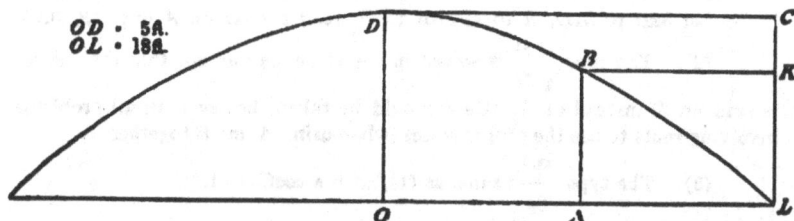

$OD : 5A.$
$OL : 18A.$

to find the distance OA which corresponds to a distance of $AB = 3.37$ feet. Subtracting 3.37 from 5 gives $1.63 = CK$. To solve, set $18C$ to $5A$, R to $1.63A$, and read answer under R on C as 10.29. This may be checked by finding the equation of the parabola, which is found to be $y^2 = \dfrac{324}{5} x$; when $x = CK$ and

$y = OA$, we have $y = \sqrt{\dfrac{324 \times 1.63}{5}} = 10.29$. Problems of this type are very

useful in structural design, where the moment curve is a parabola, or in railroads where a vertical curve is parabolic in form. The setting, as given, is deduced from the fact that on a parabola the abcissae are proportional to the squares of the ordinates. Let us work one simple problem in detail.

Given a beam uniformly loaded with a load of 1000 lbs. per foot; the beam being 17 feet long, to find the moments at one foot intervals over the span: in this case the moment curve is a parabola with vertical axis. Center distance $= \dfrac{WL^2}{8} = \dfrac{1000 \times 17^2}{8}$; $8B$ to $17D$, read the answer on A over $1B$ as 36100; set $\dfrac{17}{2}$ or $8.5C$ to $3.61A$ and corresponding to the following distances on the beam, as measured from the center, we have as per the table:

8.5	7.5	6.5	5.5	4.5	3.5	2.5	1.5	0.5
36100	28200	21200	15100	10100	6130	3130	1120	125

These values are obtained by setting as above and reading the numbers on A corresponding to distances on C. That is, $8.5C$ corresponds to $36100A$, 7.5 to $28200A$, etc. To find the moment subtract each number as found from 36100 and we have the actual moments at corresponding points.

End 1	2	3	4	5	6	7	8
0 7900	14900	21000	26000	29970	32970	34980	35980

The distances being expressed now as distances from the end of the beam instead of from the center.

XVII. SOLUTIONS OF SPECIAL FORMS.

The following solutions of general formulæ are instructive.

(1). The type $\dfrac{r^2}{gx}$; which occurs in hydraulic computations.

Let $r = 31$, $g = 32$, $x = 8$.

Mental approximation; 31 divided by 8 is approx. $= 4$.

Set $32B$ to $31D$, R to $1B$, $8B$ to R, read answer on A over $1B$, 3.75.

(2). The type $\sqrt{\dfrac{r^2}{gx}}$ is solved in the same way except that the answer is read on D instead of A. Care should be taken, however, in all problems involving roots to use the proper scales, when using A and B together.

(3). The type $\dfrac{mr^2}{2g}$; same as (1), with a coefficient.

Let $v = 8.5$, $m = .93$, $2g = 64.4$

Mental approximation: $80/64 =$ approx. 1.

Set $64.4B$ to $8.5D$, and over $.93B$ find 1.04 answer on A.

(4). The type $\dfrac{\sqrt{2gh}}{a}$;

Let $2g = 64.4$, $h = 12.5$, $a = 5$.

Mental approximation; $\dfrac{\sqrt{625}}{5} =$ approx. 5.0.

Invert the slide; set $64.4C$ to $12.5A$, under $5B$ find answer on D as 5.67. Care should be taken to use the proper scales; notice that with the wrong choice 179 is read under $5B$.

(5). The type $\dfrac{a\sqrt{b}}{c}$;

Find the square root of b, divide by c and multiply by a. This can be done either direct or inverted. Direct: set cC to bA, and under aC find the answer on D.

(6). The type $a\sqrt{\dfrac{b \times c}{d}}$; take for an example of this type the formula for the flow through a nozzle as expressed by the formula: $q = .7854d^2$ $\sqrt{\dfrac{2gh}{\left(\dfrac{1}{c}\right)^2 - \left(\dfrac{d}{D}\right)^2}}$, and let us consider its solution in detail. First find the value of $.7854d^2$, which is the area of a circle with diameter d.

Let $d = 1\tfrac{3}{4}$ in., then $.7854d^2 = 2.4$ sq. in.

Let $D = 3\tfrac{1}{2}$ in., $2g = 64.4$, $c = .97$, $h = 130$.

$\left(\dfrac{1}{c}\right)^2 = 1.06$; $\left(\dfrac{d}{D}\right)^2 = .25$ $\left[\left(\dfrac{1}{c}\right)^2 - \left(\dfrac{d}{D}\right)^2\right] = .81$. The formula has

now been reduced to $2.4\sqrt{\dfrac{2gh}{.81}}$ and this formula is of the type $\sqrt{\dfrac{b \times c}{d}}$. Mental

approximation: $65 \times 150 = 10000$, $\sqrt{10000} = 100$, $100 \times 2.4 = 240$.

Considering this formula, it is evident that when we have completed the process $\sqrt{\dfrac{b \times c}{d}}$ the result should be indicated upon scale A in order to facilitate the finding of the square root; this being found, we multiply the result by a. Any method of multiplication which gives the result on A can be used. There is no reason, therefore, why the scale should not be used direct.

Set 81B to 64.4A, R to 130B, 1B to R, under 2.4C read the answer on D as 244 cubic feet per second. Again we must be careful in selecting the scales to use on A and B. Estimating the value of the expression under the radical and mentally extracting the square root, it is seen to be about 100, therefore, the result of the quantity under the radical should appear on the left-hand scale of A. This simple approximation should always be made where a root is involved.

(7). The type $\dfrac{H^3}{b}$ or $\dfrac{\sqrt{H^3}}{b}$. The analysis shows that we should cube H so that the result falls on A, transfer to D and divide by b.

Let $H = 1.79$, $b = 3.32$. Notice that $H^3 =$ approx. 8, so we wish the result of this step to fall on LA.

Mental approximation: $3/3.3 = .9$.

Slide direct: set 1C to 179D, R to 179B, 332C to R, read answer on D under 1C, as .722. With slide inverted, set 179C to 179D, and under 332B find answer on D as .722. As the second method requires a change of index, in this particular case it is little, if any, shorter. In most cases, however, the second method is the quicker of the two, and is, therefore, to be recommended.

(8). The type $a \times b^{\frac{3}{2}}$; an example of this type is the formula $Q = 3.33bh^{\frac{3}{2}}$ due to Francis. The quantity Q is the discharge over a trapezoidal weir in cubic feet per second, b the length on the crest in feet, and h the head of the weir. The formula may be written $b/.3 \times h^{\frac{3}{2}}$; as is readily seen, the inverted slide offers the best solution.

Let $h = 1.43$, and $b = 4.51$.

Mentally approximate: $(1\frac{1}{2})^3 = 3.5$, $\sqrt{3.5} = 2$, $2 \times 5 \times 3 = 30$. Ans·

Solution with slide inverted: Set 1.43C to 1.43D, R to 1C, 4.51B to R, read answer on D under 3B as 25.7. Notice that the cube of 1.43 is approximately 3.5, so that the result of the process of cubing should appear on the left scale of A. If it is made to appear on the right, the answer is found to be 813, which is impossible.

(9). The type $\dfrac{\sqrt{a \times b}}{c}$ is sometimes given. This is, of course, equivalent to the expression $\sqrt{\dfrac{a \times b^2}{c^2}}$ and in this form, it is of more common occurrence.

Its solution requires us to find the ratio $\dfrac{b}{c}$, square it and multiply by a.

Let $a = 9.18$, $b = 3.12$, $c = 4.26$.

Mental approximation: $81/16 = 5$.

Use the slide direct; set 4.26C to 3.12D, read answer on A over 918B as 4.93. These types shown are few in number, but they serve to illustrate the method of finding a setting for a given formula.

XVIII. GENERAL CONDITIONS.

Remember in general that all formulæ should be expressed in the form of products, as the first step in their solution by means of the slide rule. The expression for the cosine of any angle of a triangle in terms of its sides is a familiar example of this process. $\text{Cos } \frac{1}{2}A = \sqrt{\dfrac{s\,(s-a)}{bc}}$, where $s = \frac{1}{2}$ the sum of the sides a, b, c. This expression is derived from the law of cosines by a simple transformation. As is readily seen, the form above is adapted to slide rule computation, while the law of cosines is not. That is, $\text{Cos } A = \dfrac{b^2 + c^2 - a^2}{2bc}$ cannot be readily solved by either the slide rule or by logarithms.

It is sometimes helpful to add and subtract numbers on the rule when paper and pencil are not at hand. This may be done by reversing and inverting the slide; suppose we have the following: Add 3.17

$$\begin{aligned} &5.36 \\ &-3.81 \\ &-7.42 \\ &6.89 \\ &-4.19 \end{aligned}$$

With the slide reversed and inverted, bring indices into coincidence, R to 3.17 *log scale*, 0 *log scale* to R, R to 5.36 *log scale*, 3.81 *log scale* to R, R to 0 *log scale*, 7.42 *log scale* to R, R to 0 *log scale*, invert slide and set right-hand indices into coincidence. Read the distance on *log scale* between runner and 6.89. Read in this way: 3 1.1 .09 or 4.19. The operator will easily grasp the principle of this setting and can readily see in what cases its application will be of advantage.

KEUFFEL & ESSER CO. NEW YORK

PROBLEMS.

EQUIVALENT RATIOS.

(1). Find the areas of circles the diameters of which are respectively: 7.93. 9.76, .01345, 64.8, and 33.71 in.

Answers: 49.4, 74.8, .0001421, 3300, 893.

(2). Find the number of cubic feet corresponding to the following numbers of U. S. gallons: 8.0, 9.78, 351, 1003.

Answers: 1.070, 1.307, 46.9, 134.1.

(3). Transform the following volumes expressed in cubic feet to litres: 33.9, 67.8, 85.4.

Answers: 960, 1921, 2420.

(4). Find decimals of inches corresponding to the following fractions of inches: 1/8, 3/32, 5/16, 9/16, 17/64, 19/64.

Answers: .125, .0938, .313, .563, .266, .297.

(5). Find decimals of a foot corresponding to the following numbers of inches: $3\frac{1}{8}$, $5\frac{1}{4}$, $7\frac{3}{4}$, $9\frac{1}{2}$, $11\frac{3}{4}$, $8\frac{3}{8}$, $4\frac{1}{2}$.

Answers: .260, .438, .646, .792, .979, .698, .375.

(6). Find the pressures in tons per square foot due to the following depths of water: $6\frac{1}{2}$, 9, 17, 31.5, and 76 feet.

Hint: Determine first the ratio between pounds per square inch and tons per square foot.

Answers: .203, .281, .531, .984, 2.38.

(7). Express to the nearest even sixteenth of an inch the following decimals of a foot: .938, .642, .733, .657.

Answers: $11\frac{1}{4}$, $7^{11}/_{16}$, $8^{13}/_{16}$, $7\frac{7}{8}$.

(8). Transform the following pressures given in inches of mercury to equivalent pressures expressed as atmospheres: 35.7, 33.4, 31.6, 27.3.

Answers: 1.194, 1.117, 1.057, .910.

(9). Find the weight of copper strips 1 in. wide, 1 ft. long, and of the following thicknesses: $^3/_{16}$, $\frac{1}{4}$, $\frac{5}{8}$, $1\frac{1}{8}$ in.

Answers: .716, .953, 2.38, 4.28.

(10). Find the weight in grams of a steel plate circular in shape with diameter .1125 in. and thickness $^{11}/_{32}$ of an inch.

Answer: .759.

SQUARES AND SQUARE ROOTS.

(1). Find the squares of the following: 10.17, .979, 126.5, .0326, 4.77, 87.9.

Answers: 103.4, .958, 16000, .001063, 22.75, 7730.

(2). Find the square roots of the following: .0735, .520, 1391, 600, 3.17.

Answers: .271, .721, 37.3, 24.5, 1.78.

(3). Find the squares of the following: 43/52, 47/64, 89/71.

Hint: 52C to 43D, read answer on A over 1B.

Answers: $.683$ or $\dfrac{24}{35}$ or $\dfrac{8.9}{13}$; $.540$ or $\dfrac{27}{50}$ or $\dfrac{34.5}{64}$; 1.57 or $\dfrac{110}{70}$ or $\dfrac{100.5}{64}$.

(4). Find the square roots of the following expressed as even sixty-fourths: 9/64, 19/32, 3/8, 7/16, 1/32.

Answers: 24/64, 49/64, 39/64, 42/64, 11/64.

(5). Find the length of the side of a square to the nearest 64th of an inch, the square containing 9.81 square inches.

Answer: $\dfrac{200}{64}$ or $3\frac{1}{8}$.

(6). Find the diameter of a circle to the nearest thirty-second of an inch, the circle containing 7.97 square inches.

Hint: Use the ratio found on the reverse of the rule 79:70.

Answer: $\dfrac{204}{64}$ or $3^3/_{16}$.

MULTIPLICATION AND DIVISION.

(1). Solve $16.16 \times 3.1416 \times 18.13$. Answer: 921.

(2). Solve $\dfrac{39.8}{46.7}$. Answer: .852.

(3). Solve $\dfrac{83.7 \times 11.63}{77.6}$. Answer: 12.54.

(4). Solve $\dfrac{100.3 \times 69.3 \times 33.7}{88.2 \times 54.6 \times 17.70}$. Answer: 2.75.

(5). Solve $\dfrac{1131 \times 94.8}{79.9 \times 104.4 \times 73.5}$. Answer: .1750.

(6). Solve $\dfrac{24}{\dfrac{\dfrac{47}{\dfrac{18}{\dfrac{51}{\dfrac{1}{19}}}}}{}}$. Answer: 0.0761.

√ (7). Solve the following proportion, stating the answer to the nearest thirty-second: $\dfrac{17\frac{1}{4}}{3\frac{1}{4}} = \dfrac{9\frac{3}{4}}{x}$. *Answer:* $x = \dfrac{54}{32}$ or $1\dfrac{22}{32}$ or $1\dfrac{11}{16}$.

(8). Solve the following: $\dfrac{(3.18)^2}{(10.17)^2} = \dfrac{x}{33.4}$. *Answer:* $x = 3.27$.

(9). Give the general setting for the solution of problems of the type of (8) $\dfrac{a^2}{b^2} = \dfrac{x}{c}$. *Answer:* Set bC to aD, answer on A over cB.

(10). Find the answer to this problem in kilogram meters when the quantities are expressed in feet and pounds:

$$\frac{3.1416 \times (5.07)^2 \times 189 \times 2.73 \times 550}{4}.$$

Answer: 791000.

Hint: Deduce from the table of equivalents a factor to satisfy this case and multiply the expression by it.

INVERTED SLIDE.

(1). Find the reciprocals of: 389, 40.9, 7.79, 1019, 6.59.

Answers: .00257, .0244, .1284, .000981, .1517.

(2). Multiply 13.14 × 9.63 × 7.46 × 349. *Answer:* 329000.
Multiply 159.6 × .834 × 76.7 × 9.13. *Answer:* 93200.

(3). Solve this example with the slide inverted and also with the slide direct:

$$\frac{3.43 \times 79.6 \times 928 \times .1119 \times .06888}{19.15 \times 47.7 \times 344 \times 42.2}.$$

Answer: .0001471.

(4). Find a simple method for solving the following and determine their numerical values: $\dfrac{1}{(3.29)^2}, \dfrac{1}{(18.72)^2}, \dfrac{1}{(4.13)^2}, \dfrac{1}{(1.251)^2}$

Answer: Invert slide and read over 3.29, 18.72, etc., on B, the answer on A as: .0924, .00285, .0586, .638.

(5). Find a simple setting for problems of this class and determine numerical results with the data as given.

A lever of the first class has the fulcrum 6 in. from the end where rests a weight of 380 lbs. At what distances from the fulcrum will the weights given be placed, one weight acting at a time, in order to insure equilibrium? The weights are 50, 73.4, 16.93, and 217 lbs.

Answer: Invert slide, set $5B$ to $380D$ and under the number of pounds in B find the answer in feet on D as follows: 3.80, 2.59, 11.22, .875.

(6). Find the supporting forces on the two ends of a beam 22.3 feet long under the action of a movable load of 625 pounds. Reactions are desired when the load is concentrated at distances of 2, 4, 6, 8, 10, 12 feet from the left-hand end of the beam. Answers to be given to the nearest pound. *Answers* 55–570, 112–513, 168–457, 224–401, etc.

(7). Find the number of amperes required to run a motor developing 15 horsepower at voltages of 110, 220, 550, and 750; using the formula H. P.—$\frac{V \times A}{746}$. *Answer:* 102, 51, 20.4, 14.9.

(8). At a pressure of two atmospheres a certain quantity of a certain gas has a volume of 22 cubic feet. Assuming Boyle's law to hold, that is: PRESSURE $\times$ VOLUME = CONSTANT, find the volume of the gas at pressures of 33, 37, 40, 45, 46 pounds per square inch.

Answers: 19.6, 17.5, 16.2, 14.4, 14.1.

(9). If a pump A of capacity 47 cubic feet per minute can empty an excavation in $12\frac{1}{4}$ days at a cost of $3.45 per day, and if a pump B, capacity 38 cubic feet per minute, which runs at a cost of $2.83, is also available, which will be the cheaper to use? *Answer:* A.

(10). State the simplest setting for (9).

Answer: Invert slide. Set 47B to 1225D, R to 38B, 283B to R, R to 1225D. If the number on B under R is greater than 345, A is the cheaper.

THE FORMS $XY = B$ AND $X^2Y = B$.

(1). Find the dimensions of a rectangle, its sides being determined to the nearest inch only, so that the rectangle shall contain as nearly as possible 1317 square inches. The short side of the rectangle must be over 25 in long. *Answer:* 28 $\times$ 47 in.

(2). Determine the dimensions of a right triangle so it will contain, at least, and as near 317 square inches as possible. The sides are to be measured to the nearest $\frac{1}{2}$ in. The short leg of the triangle must be 15 in. or over. *Answer:* $29\frac{1}{2} \times 21\frac{1}{2}$ in.

(3). A sewer must have a cross sectional area of at least 716 square inches and is elliptical in section. Find the lengths of the major and minor axes to the nearest inch. The minor axis must be between 20 and 25 inches long. Area of ellipse is .7854 $a \times b$, where a is the major and b the minor axis. *Answer:* 24 $\times$ 38.

(4). Determine the number, width and thickness of steel plates for the flange of a girder, so that the gross cross-sectional area of the plates shall exceed by an amount as small as possible, 23.7 square inches. The widths of the plates can vary by half inches from 10 to 12 inches, and in thickness from $\frac{3}{8}$ to $\frac{5}{8}$ by sixteenths. The plates must all be the same width and should vary in thickness as little as possible.

Answer: 3 Pl. $11\frac{1}{2} \times \frac{1}{2}$ in., 1 Pl. $11\frac{1}{2} \times \frac{9}{16}$ in.

(5). Determine plates required under conditions which are identical with problem (4), except that the gross area must be 43.6 and the plates may vary in width by half inches from 14 to 18 in.

Answer: 5 Pl. 14 $\times \frac{5}{8}$ in., or 4 Pl. $17\frac{1}{2} \times \frac{5}{8}$ in.

KEUFFEL & ESSER CO., NEW YORK

(6). Design the section of a rectangular wooden beam using the formula $M = \dfrac{f b h^2}{6}$, where M = 31250 inch pounds, and f = 980 lbs. per square inch. The minimum value of b is two inches. Both b and h must be multiples of two. *Answer:* 2 × 10 in.

(7). Design a beam under the same conditions as problem (6) where M = 72500 inch pounds and f = 1000 lbs. per square inch. Minimum value for b is 6 in. *Answer:* 6 × 10 in.

(8). Determine b and h for a rectangular reinforced concrete beam where $M = 125\, b\, h^2$. M = 1500000 inch pounds. h must be at least double b, and b at least 13 in. Determine the dimensions to the nearest ½ in. *Answer:* 13½ x 30 in.

CUBES.

(1). Cube the following numbers: 728, 8.07, 55.9, 10.17.
Answers: 386000000, 526, 174700, 1052.

(2). Cube the following with the slide inverted: 9.77, 100.5, .246, .0133. *Answers:* 933, 1015000, .01489, .000002353.

(3). Solve the following with the slide direct: $(15.1)^3$, $(2.34)^3$, $(.0327)^3$. $(^{13}/_{11})^3$. *Answers:* 58.7, 3.58, .00591, $^{17}/_{12}$ or .532.

(4). Deduce the setting for x^3 with the slide inverted, when x lies between 1 and 10. *Answer:* xRC to xD, under $1RC$ read answer on D, or xB to xLA, under $1RC$ read answer on D.

(5). With the scale inverted find the following; expressing the answer to the nearest thirty-second of an inch. $(3.71)^3$, $(^7/_{16})^3$, $(1^1/_8)^3$, $(47.5)^3$.

$$Answers \quad \frac{229}{32}, \frac{9}{32}, \frac{38}{32}, \frac{10480}{32}.$$

CUBE ROOTS.

(1). Find the cube roots of the following with the slide direct: 2.613, 17.8 .00321, 1793, .0488. *Answers:* 1.377, 2.61, .1475, 12.15, .3655.

(2). Find the cube roots of the following with the slide inverted: 8.93, 14.16, .000444, .01331, 571. *Answers:* 1.578, 2.42, .0763, .237, 8.30.

(3). Solve the following: $\sqrt[3]{(8.93)^2}$, $\sqrt[3]{(71.7)^2}$, $\sqrt[3]{(1.423)^2}$, $\sqrt[3]{(.0321)^2}$. *Answers:* 4.30, 17.26, 1.265, .1010.

(4). Deduce the general setting for the type H^3, with the slide inverted. *Answer:* $1C$ to IID. Look for coincidences between A and B, or C and D. The numbers coinciding will give the answer.

(5). Solve the equation $x^{\frac{3}{2}}$ = 18.17. *Answer:* x = 77.5.

(6). Deduce the setting for the solution of the type shown in (5).
Answer: Set right-hand member on A to right-hand member on B and read x on D under $1C$. Slide inverted.

(7). Solve the equation $x^{\frac{3}{2}} = 9.81$. *Answer:* $x = 4.58$.

(8). Deduce the setting for the solution of the type shown in (7) with the slide direct. *Answer:* R to right-hand member on D, move the slide until the same number on B is under R, that appears on D under $1C$. This number will be the required value of x.

(9). A right circular cone contains 347 cubic inches and the radius of the base is $7\frac{1}{4}$ in. What will be the volume of a similar cone, the radius of the base being 8.19 in.? *Answer:* 500 cu. in.

(10). If a sphere of (a) inches radius contains 400 gallons, what will be the radius of a sphere containing 600 gallons. *Answer:* 1.144 a.

(11). Find in problem (10) the radii of the two spheres in feet. *Answer:* 2.33 ft. and 2.67 ft.

LOGARITHMS, INVOLUTION AND EVOLUTION.

(1). Find the logarithms of the following numbers to three places in the mantissa: 2181, 131.7, .983, .0436, 7.94.
Answers: 3.339, 2.120, 9.993−10, 8.639−10, 0.900.

(2). Find the numbers corresponding to the following logarithms: 2.736, 7.1316−10, .4717, 6.979−10. *Answers:* 544.5, .001354, .2963, .000953.

(3). Find the fifth power of 3.72 by two distinct methods.
Answer: 712.

(4). Solve $(1.217)^{\frac{3}{5}}$ by the use of logarithms. *Answer:* 1.387.

(5). Solve $(8.44)^{1.77}$ and $(.00511)^{3.11}$. *Answers:* 15.01 and .0000000747.

(6). Solve $(.0132)^{-.405}$ and $\sqrt[-21]{1729}$. *Answers:* .173 and 27.8 billions.

EQUATIONS INVOLVING FRACTIONAL POWERS.

(1). Solve the equation $S = \dfrac{W e^{\frac{wx}{f}}}{f}$, when W is 10 tons, $x = 100$ feet, $w = .27$ pounds per cubic inch, $f = 10000$ lbs. per square inch. Value for e will be found on the reverse of the rule. Express the answer in square inches. *Answer:* 2.066.

(2). Using Hodgkinson's formula $W = \dfrac{99318 \times D^{3.55}}{L^{1.7}}$, find W when $D = 2.5$ and $L = 17.8$. *Answer:* 19230 lbs. per sq. in.

(3). Solve the equation $x^{2.15} = 3.43$. *Answer:* $x = 1.760$.

(4). Solve for t: $t = \dfrac{14 \times 3.1416 \times (5)^{\frac{3}{2}}}{\dfrac{15}{4}\sqrt{64.4 \times 12}}$. *Answer:* $t = 23.6$.

(5). Solve for q in the equation: $q = 3.01 \times 16.7 \times (2.32)^{1.90}$
Answer: 182.2

TRIGONOMETRIC COMPUTATIONS.

(1). Find sin 32° 14′, sin 85° 38′, tan 19° 14′, tan 22° 41′.
Answers: .533, .997, .349, .418.

(2). Find cos 29° 17′, cos 73° 04′, cot 89° 19′, cot 12° 51′.
Answers: .872, .291, .01193, 4.38.

(3). Find sec 45° 09′, sec 73° 33′, csc 11° 13′, csc 36° 28′.
Answers: 1.419, 3.53, 5.14, 1.685.

(4). Find sin 00° 18′, tan 00° 07′, sin 00° 41′, tan 01° 03′.
Answers: .00524, .00204, .01193, .01833.

(5). Perform the following operations:

$$86.1 \times \sin 30° 35', \quad \frac{\sin 41° 03' \times 386}{\sin 10° 17'}, \quad \frac{\tan 20° 10' \times 31.3}{\tan 43° 16'}.$$

Answers: 18.4, 1420, 12.21.

(6). Find log sin 39° 19′, log sin 15° 52′, log tan 40° 40′, log cot 49° 35′, log sec 86° 17′, log csc 4° 44′.
Answers: 9.802–10, 9.437–10, 9.934–10, 9.930–10, 1.188, 1.083.

(7). Solve the triangle as given.

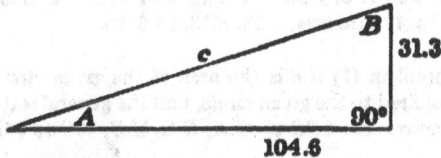

Answers: A = 16° 39′, B = 73° 21′, c = 109.2.

(8). Solve the triangle as given.

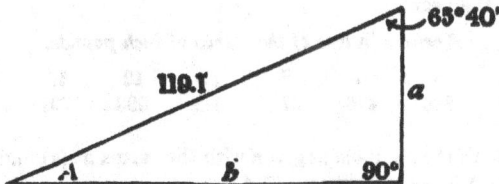

Answers: A = 24° 20′, b = 108.5, a = 49.1.

(9). Solve the triangle as given.

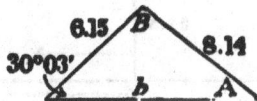

Answers: A = 22° 14′, B = 127° 43′, b = 12.86.

(10. Solve the triangle as given.

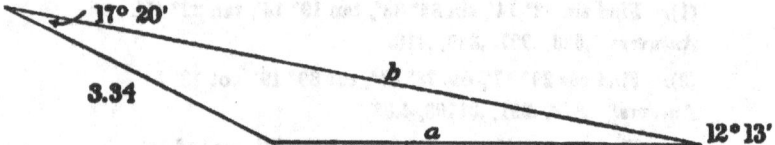

$\angle$ 17° 20'

3.34

b

a

12° 13'

Answers: $B = 150° 27'$, $b = 7.76$, $a = 4.70$.

VARIATION AS THE SQUARE.

(1). A circle has an area of 34.71 square inches. What will be the diameter of a circle whose area is $3\frac{1}{4}$ times as large? *Answer:* 12 in.

(2). An equilateral triangle has its sides 3.17 in. long and has an area of 4.33 sq. in. What is the area of an equilateral triangle whose side is 3.47 in.; one whose side is 7.62; one whose side is 17.1 ?
   Answers: 5.18, 25.01, 125.9 in.

(3). A bar of steel 1 ft. long and $1\frac{1}{4}$ in. in diameter weighs 4.17 lbs. What will be the weight of bars 1 ft. long and with the following diameters: $1\frac{1}{8}$, $1\frac{3}{4}$ and $2\frac{1}{2}$ in. ? *Answers:* 3.38, 8.18, 16.6 lbs.

(4). In problem (1) if c is the area of the given circle, b the ratio of the area of the required to the given circle, find the general setting for problems of this type. *Answer:* Set 1RB to cRA, R to bLB, 79C to 70D, answer under runner on C.

(5). A beam 28 feet long is loaded with a uniform load such that the moment at the center is 7070000 inch pounds. Find the moments at 2 foot intervals over the beam.

Answers in tens of thousands of inch pounds.

0	2	4	6	8	10	12	14	16 etc.
0	188	346	476	577	649	693	707	693

(6). With the parabola as given with the vertex at (a) find the distance (b) so that c is 21.5 feet. *Answer:* 61.4 ft.

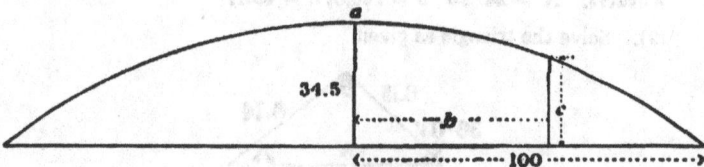

34.5

a

b

c

100

KEUFFEL & ESSER CO., NEW YORK

(7). A parabola as given with the vertex at a; $dc = 238'$, $ad = 440'$. Determine perpendicular offsets to the parabola from points 40 feet apart

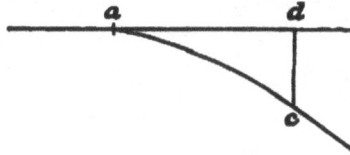

on ad *Answer:*

a	40	80	120	160	200	240	280	320	360	400
0	1.9	7.9	17.7	31.5	49.3	71.0	96.8	126.0	160	197

SPECIAL FORMS.

(1). Compute $\dfrac{(8.72)^2}{32.2 \times 8.17}$. *Answer:* .289.

(2). Compute $\sqrt{\dfrac{(91.5)^2}{32.2 \times 17.17}}$. *Answer:* 3.90.

(3). Compute $\dfrac{.925 \times (13.11)^2}{64.4}$. *Answer:* 2.47.

(4). Compute $\dfrac{\sqrt{64.4} \times \sqrt{73.8}}{4.36}$. *Answer:* 15.81.

(5). Compute $\dfrac{13.16 \times \sqrt{.036}}{19.37}$. *Answer:* .1290.

(6). Compute $8.88 \sqrt{\dfrac{2.718 \times 3.4}{73.7}}$. *Answer:* 3.14.

(7). Compute $\dfrac{\sqrt{(3.13)^2}}{13.19}$. *Answer:* .420.

(8). Compute $8.9 (.032)^{\frac{1}{2}}$. *Answer:* .0510.

(9). Compute $\dfrac{\sqrt{3.18} \times 9.72}{1.331}$. *Answer:* 13.02.

(10). Compute $\sqrt{\dfrac{32.9 \times (1.313)^2}{(.977)^2}}$. *Answer:* 7.71.